ANNALS OF THE NEW YORK ACADEMY OF SCIENCES

Volume 510

EDITORIAL STAFF

Executive Editor
BILL BOLAND

Managing Editor
JUSTINE CULLINAN

Associate Editor
LINDA H. MEHTA

The New York Academy of Sciences
2 East 63rd Street
New York, New York 10021

OLFACTION AND TASTE

ANNALS OF THE NEW YORK ACADEMY OF SCIENCES

Volume 510

OLFACTION AND TASTE IX

Edited by Stephen D. Roper and Jelle Atema

The New York Academy of Sciences
New York, New York
1987

Library of Congress Cataloging-in-Publication Data

Main entry under title:

Olfaction and taste IX.

  (Annals of the New York Academy of Sciences ;
v. 510)
  Based on the Ninth International Symposium on Olfaction
and Taste held at the Snowmass Village, Colorado, July
20-24, 1986 by the International Commission on Olfaction
and Taste.
  Includes bibliographies and index.
  1. Smell—Physiological aspects—Congresses.
2. Taste—Physiological aspects—Congresses.
I. Roper, Stephen D.   II. Atema, Jelle.
III. International Symposium on Olfaction and
Taste (9th : 1986 : Snowmass Village, Colo.)
IV. International Commission on Olfaction and Taste.
Q11.N5 vol. 510   500 s   [612'.86]   87-34753
[QP458]
ISBN 0-89766-416-7
ISBN 0-89766-415-9 (pbk.)

PCP
Printed in the United States of America
ISBN 0-89766-415-9 (paper)
ISBN 0-89766-416-7 (cloth)

ANNALS OF THE NEW YORK ACADEMY OF SCIENCES

Volume 510
November 30, 1987

OLFACTION AND TASTE IX [a]

Editors
STEPHEN D. ROPER AND JELLE ATEMA

CONTENTS

Preface. *By* STEPHEN D. ROPER AND JELLE ATEMA................. xxi

Part I. Symposium Papers

Evolutionary Patterns in Sensory Receptors: An Exercise in
Ultrastructural Paleontology. *By* DAVID T. MORAN............. 1

Sensory Transduction in Flagellate Bacteria. *By* JUDITH P.
ARMITAGE AND R. ELIZABETH SOCKETT 9

Eukaryotic Unicells: How Useful in Studying Chemoreception? *By*
JUDITH VAN HOUTEN AND ROBIN R. PRESTON 16

Ionic Mechanism of Generation of Receptor Potential in Frog Taste
Cells. *By* TOSHIHIDE SATO, YUKIO OKADA, AND TAKENORI
MIYAMOTO.. 23

Toward a Comprehensive Molecular Analysis of Olfactory
Transduction. *By* DORON LANCET, ZEHAVA CHEN, ADINA
CIOBOTARIU, FRITZ ECKSTEIN, MIRIAM KHEN, JUDITH
HELDMAN, DOV OPHIR, IRIS SHAFIR, AND UMBERTO PACE ... 27

Neural Derivation of Sound Source Location in the Barn Owl: An
Example of a Computational Map. *By* ERIC I. KNUDSEN 33

Neurophysiology and Neuroanatomy of the Olfactory Pathway in
the Cockroach. *By* J. BOECKH, K.-D. ERNST, AND P. SELSAM 39

Organization of Olfactory Bulb Output Cells and Their Local
Circuits. *By* JOHN W. SCOTT 44

Cortical Organization in Gustatory Perception. *By* TAKASHI
YAMAMOTO 49

Effects of Odorant Mixtures on Olfactory Receptor Cells. *By*
STEVEN PRICE 55

[a] This volume is the result of a symposium entitled Ninth International Symposium on Olfaction and Taste, held in Snowmass Village, Colorado on July 20-24, 1986, by the International Commission on Olfaction and Taste.

Coding of Chemosensory Stimulus Mixtures. *By* DAVID G. LAING .. 61

Psychophysical Models for Mixtures of Tastants and Mixtures of Odorants. *By* JAN E. R. FRIJTERS 67

Electrophysiological Responses of Olfactory Receptor Neurons to Stimulation with Mixtures of Individual Pheromone Components. *By* ROBERT J. O'CONNELL AND ALAN J. GRANT 79

Odor-Taste Mixtures. *By* DAVID E. HORNUNG AND MELVIN P. ENNS 86

The Mixture Symposium: Summary and Perspectives. *By* MAXWELL M. MOZELL... 91

How Motile Bacteria Sense and Respond to Chemicals. *By* JULIUS ADLER.. 95

A Molecular Vocabulary for Olfaction. *By* GORDON M. SHEPHERD 98

Adaptation Processes in Insect Olfactory Receptors: Mechanisms and Behavioral Significance. *By* K.-E. KAISSLING, C. ZACK STRAUSFELD, AND E. R. RUMBO.............................. 104

A Standardized Olfactometer in Japan: A Review Over Ten Years. *By* SADAYUKI F. TAKAGI....................................... 113

Visual Approach to Fragrance Description. *By* MAURICE THIBOUD 119

Part II.   Workshop Summaries

Trigeminal Chemoreception. DANIEL KURTZ, *Chair* 127

Temporal Aspects of Chemical Stimuli: Natural Stimulus Distributions, Receptor Cell Adaptation, and Behavioral Function. JELLE ATEMA, Chair 130

How Is Peripheral Input Processed in the Central Nervous System? JOHN SCOTT, *Chair* 133

Seeking the Mechanisms of Chemical Transduction. VINCENT E. DIONNE, *Chair*... 135

Salt Appetite: Intake and Preference for Salt in Man and Animals. MARY BERTINO, *Chair* 137

Recent Trends in Clinical Measurement of Taste and Smell. JANNEANE F. GENT, *Chair*..................................... 140

Part III.   Poster Papers

Mechanism of Interaction between Odorants at Olfactory Receptor Cells. *By* BARRY W. ACHE, RICHARD A. GLEESON, AND HOLLY D. THOMPSON 145

Reproductive Pheromones of the Landlocked Sea Lamprey *(Petromyzon marinus):* Studies on Urinary Steroids. *By*

MICHAEL A. ADAMS, JOHN H. TEETER, YAIR KATZ, AND
PETER B. JOHNSON 148

Transduction Proteins of Olfactory Receptor Cells: Identification of
Guanine Nucleotide Binding Proteins and Protein Kinase C.
By ROBERT R. H. ANHOLT, SUSANNE M. MUMBY, DORIS A.
STOFFERS, PEGGY R. GIRARD, J. F. KUO, A. G. GILMAN,
AND SOLOMON H. SNYDER 152

Granule Cell Development in the Ferret Olfactory Bulb under
Normal Conditions and under Continuous Overexposure to a
Single Odor. *By* R. APFELBACH AND E. WEILER 157

Morphological and Physiological Characterization of Interneurons
in the Olfactory Midbrain of the Crayfish. *By* EDMUND A.
ARBAS, CAROL J. HUMPHREYS, AND BARRY W. ACHE 160

Dopamine and Substance P Are Contained in Different Populations
of Tufted Cells in the Syrian and Chinese Hamster Main
Olfactory Bulb. *By* HARRIET BAKER 164

Tasting on Localized Areas. *By* LINDA BARTOSHUK, SALLI
DESNOYERS, COURTNEY HUDSON, LAURA MARKS,
MARGARET O'BRIEN, FRANK CATALANOTTO, JANNEANE
GENT, DORI WILLIAMS, AND KAREN M. OSTRUM 166

Effects of Vomeronasal Organ Removal in Lactating Female Mice:
Dissociation of Maternal and Agonistic Behaviors. *By* N. JAY
BEAN AND CHARLES J. WYSOCKI 169

Preference for Extremely High Levels of Salt among Young
Children. *By* GARY K. BEAUCHAMP AND BEVERLY J.
COWART.. 171

Schematic Sweet and Bitter Receptors. *By* HANS DIETER BELITZ,
HARTMUT ROHSE, WOLFGANG STEMPFL, HERBERT WIESER,
JOHANN GASTEIGER, AND CHRISTIAN HILLER 173

Early-Stage Processing of Odor Mixtures. *By* GRAHAM A. BELL,
DAVID G. LAING, AND HELMUT PANHUBER................... 176

Concentration-Dependent Responses of Sodium-Depleted Rats to
NaCl in Food. *By* MARY BERTINO, MICHAEL G. TORDOFF,
AND JOHN TLUCZEK.. 178

The Structure of Chemosensory Centers in the Brain of Spiny
Lobsters and Crayfish. *By* DAVID BLAUSTEIN, CHARLES D.
DERBY, AND ARTHUR C. BEALL............................... 180

Self- and Cross-Adaptation of Single Chemoreceptor Cells in the
Taste Organs of the Lobster *Homarus americanus*. *By* PAOLA
F. BORRONI AND JELLE ATEMA 184

Does the Trigeminal Nerve Control the Activity of the Olfactory
Receptor Cells? *By* JEAN-FRANCOIS BOUVET, JEAN-CLAUDE
DELALEU, AND ANDRE HOLLEY............................... 187

Feeding Regulation and the Microstructure of Eating Tomato Leaf
by Tobacco Hornworm Caterpillars. *By* ELIZABETH BOWDAN   190

Enantiomeric Specificity of Alanine Taste Receptor Sites in Catfish. *By* JOSEPH G. BRAND, BRUCE P. BRYANT, ROBERT H. CAGAN, AND D. LYNN KALINOSKI 193

Behavioral Reactions to Taste Stimuli in Hatchling Chicks. *By* A. BRAUN-BARTANA, J. R. GANCHROW, AND J. E. STEINER 196

Taste Detection and Discrimination in Zinc-Deprived Rats. *By* G. M. BROSVIC, B. M. SLOTNICK, AND R. I. HENKIN.............. 199

Olfactory Recognition of Congenic Strains of Rats. *By* R. E. BROWN, P. B. SINGH, AND B. ROSER.......................... 202

Odorant- and Guanine Nucleotide-Stimulated Phosphoinositide Turnover in Olfactory Cilia. *By* RICHARD C. BRUCH AND TAUFIQUL HUQUE 205

Use of Monoclonal Antibodies to Characterize Amino Acid Taste Receptors in Catfish: Effects on Binding and Neural Responses. *By* BRUCE P. BRYANT, JOSEPH G. BRAND, D. LYNN KALINOSKI, RICHARD C. BRUCH, AND ROBERT H. CAGAN 208

Time-Intensity Analysis of Gustatory Stimuli: Preliminary Assessment of a New Technique. *By* DARLENE BURKE, ALIKI AKONTIDOU, AND ROBERT A. FRANK.......................... 210

Influence of Aging on Recognition Memory for Odors and Graphic Stimuli. *By* WILLIAM S. CAIN AND CLAIRE L. MURPHY 212

Prediction of Olfactory Responses to Stimulus Mixtures by Cross-Adaptation Experiments. *By* JOHN CAPRIO, JOHN DUDEK, AND JESSE J. ROBINSON II 216

ATP-Sensitive Olfactory Receptors: Similarities to P_2-Type Purinoceptors. *By* WILLIAM E. S. CARR, RICHARD A. GLEESON, BARRY W. ACHE, AND MARSHA L. MILSTEAD...... 219

Taste Perception of Salt in Young, Old, and Very Old Adults. *By* J. CHAUHAN, Z. J. HAWRYSH, C. KO, AND S. KO 222

Pheromonal Information Coding by Projection Neurons in the Antennal Lobes of the Sphinx Moth *Manduca sexta. By* THOMAS A. CHRISTENSEN AND JOHN G. HILDEBRAND 224

Experiential and Endocrine Dependence of Gonadotropin Responses in Male Mice to Conspecific Urine. *By* ANDREW N. CLANCY, F. H. BRONSON, A. G. SINGER, W. C. AGOSTA, AND F. MACRIDES.. 227

Olfactory Discrimination of Plant Volatiles by the European Starling. *By* LARRY CLARK AND J. RUSSELL MASON 229

Olfactory Input to the Prefrontal Cortex in the Rat. *By* MARIE-CHRISTINE CLUGNET AND JOSEPH L. PRICE 231

Ventilatory Response of the Tegu Lizard to Upper Airway Hypoxia and Hyperoxia during Normocapnia or Hypercapnia. *By* E. L. COATES, T. M. CATON, AND G. O. BALLAM 236

Salt Deprivation and Amiloride-Induced Alterations in Neural Gustatory Responses Predict Salt Intake in Sham Drinking Rats. *By* ROBERT J. CONTRERAS................................. 239

Neurosurgical Applications of Clinical Olfactory Assessment. *By* RICHARD M. COSTANZO, PETER G. HEYWOOD, JOHN D. WARD, AND HAROLD F. YOUNG.............................. 242

Perceived Burn and Taste Intensity of Physical Mixtures of Capsaicin and Taste Stimuli. *By* BEVERLY J. COWART.......... 245

Scanning and High-Voltage Electron Microscopy of Taste Cells in the Mudpuppy, *Necturus malculosus. By* T. A. CUMMINGS, R. J. DELAY, AND S. D. ROPER.................................... 247

Purinergic Receptors Occur Externally on the Olfactory Organs and Internally in the Brain of the Spiny Lobster. *By* CHARLES D. DERBY, WILLIAM E. S. CARR, AND BARRY W. ACHE.......... 250

Off Responses to Gustatory Stimuli in the Parabrachial Pons of Decerebrate Rats. *By* PATRICIA M. DI LORENZO............... 254

Membrane Conductance Mechanisms of Dissociated Cells from the Olfactory Epithelium of the Mudpuppy, *Necturus maculosus. By* VINCENT E. DIONNE.. 258

Olfactory Dysfunction in Alzheimer's Disease: A Summary of Recent Findings. *By* RICHARD L. DOTY AND PATRICIO REYES 260

Rats Eating Together Prefer the Taste of Their Food. *By* HEATHER J. DUNCAN, AUDREY BUXBAUM, AND MICHAEL G. TORDOFF 263

A Psychophysical-Decision Model for Sensory Difference Detection Methods. *By* DANIEL M. ENNIS AND KENNETH MULLEN...... 265

Contributions of Smell and Taste to the Pleasantness of Flavor. *By* MELVIN P. ENNS AND DAVID E. HORNUNG................... 268

Immunofluorescent Studies of the Development of Rat Olfactory Epithelium. *By* A. I. FARBMAN, V. McM. CARR, J. I. MORGAN, AND J. L. HEMPSTEAD.............................. 271

Consistency of Preferences for Salt in Different Foods. *By* RICHARD SHEPHERD AND CYNTHIA A. FARLEIGH....................... 273

Human Taste Detection Thresholds: Each Subject's Threshold for a Single Stimulus Is Unique. *By* A. FAURION, T. LARDIER, J. X. GUINARD, AND B. NAUDIN 276

Chemosensory Discrimination: Behavioral Abilities of the Spiny Lobster. *By* JACQUELINE B. FINE-LEVY, CHARLES D. DERBY, AND PETER C. DANIEL.. 280

Immunoreactivity to Neuronal Growth-Dependent Membrane Glycoprotein Occurs in a Subset of Taste Receptor Cells in Rat Taste Buds. *By* THOMAS E. FINGER, HEMA SRIDHAR, MARY WOMBLE, VAR L. ST. JEOR, AND JOHN C. KINNAMON 284

Electrophysiological Basis of the Response of Olfactory Receptors to Odorant and Current Stimuli. *By* STUART FIRESTEIN AND FRANK WERBLIN 287

The Suppressed Response of NaCl following Amiloride: A Halogen-Specific Effect. *By* BRADLEY K. FORMAKER AND DAVID L. HILL.. 290

Analysis of Tastes in Mixtures by Hamsters. *By* MARION E. FRANK, SHILPA J. PATEL, AND THOMAS P. HETTINGER 293

Mixture Integration in Heterogeneous Taste Quality Mixtures: An Assessment of Subadditivity for Total Mixture Intensity. *By* ROBERT A. FRANK AND GARY ARCHAMBO..................... 296

Differential Effects of Cooling on the Intensity of Taste. *By* SANDRA P. FRANKMANN AND BARRY G. GREEN.............. 300

How the Olfactory System Generates its Chaotic Background "Spontaneous" Electroencephalographic and Unit Activity. *By* WALTER J. FREEMAN 304

Development of Taste Buds in the Chick Embryo (*Gallus gallus domesticus*). By J. R. GANCHROW AND D. GANCHROW........ 307

Taste Signal Strengths of Salt and Water Stimuli Provide a Model for Sensory Difference Tests, Involving Adaptation, Learning, and Variation in Supra- and Subadapting Sensitivity. *By* L. GOLDSTEIN, N. ODBERT, U. FARELLE, AND M. O'MAHONY... 309

A Comparative Study of the Neurophysiological Response Characteristics of Olfactory Receptor Neurons in Two Species of Noctuid Moths. *By* ALAN J. GRANT, ROBERT J. O'CONNELL, AND ABNER M. HAMMOND, JR.................... 311

The Sensitivity of the Tongue to Ethanol. *By* BARRY G. GREEN 315

Conjugate Internalization of Apposed Dendritic Membranes During Synaptic Reorganization in the Olfactory Bulbs of Adult PCD Mice. *By* CHARLES A. GREER................................... 318

Taste Dysfunction in Burning Mouth Syndrome. *By* M. GRUSHKA, B. J. SESSLE, AND T. P. HOWLEY.............................. 321

Beta-Adrenergic Control of von Ebner's Glands in the Rat. *By* SUAT GURKAN AND ROBERT M. BRADLEY 325

Earthworm Alarm Pheromone Is a Garter Snake Chemoattractant. *By* JEFFREY HALPERN, NANCY SCHULMAN, AND MIMI HALPERN.. 328

Garter Snake Response to the Chemoattractant in Earthworm Alarm Pheromone Is Mediated by the Vomeronasal System. *By* NANCY SCHULMAN, EVELYN ERICHSEN, AND MIMI HALPERN 330

Latencies of Synaptic Potentials in Odor Responses of Salamander Mitral-Tufted Cells. *By* K. A. HAMILTON AND J. S. KAUER ... 332

Monoclonal Antibodies to Multiple Glutathione Receptors Mediating the Feeding Response of *Hydra*. *By* KAZUMITSU

Hanai, Masahiko Sakaguchi, Sachiko Matsuhashi, Katsuji Hori, and Hiromichi Morita 335

Taste Responsiveness of Hamster Glossopharyngeal Nerve Fibers. *By* Takamitsu Hanamore, Inglis J. Miller, Jr., and David V. Smith.. 338

Effects of Chemical Noise on the Detection of Chemical Stimuli. *By* Linda S. Handrich and Jelle Atema 342

Neural and Behavioral Responses to Amino Acids in Mice and Rats. *By* S. Harada, T. Marui, and Y. Kasahara 345

Quantitative Analysis of Feeding Behavior Stereotypes and of the Rejection of Aversive-Tasting Food in the Freshwater Prawn *Macrobrachium rosenbergii. By* Sheenan Harpaz and Jacob E. Steiner.. 347

A Diffusion Potential Model of Salt Taste Receptors. *By* Harry Wms. Harper 349

Analysis of Olfactory Neural Responses by a Method of Spike Train Matching. *By* T. A. Harrison, J. W. Scott, and F. H. Schmidt 352

Effects of Cadmium on the Rat Olfactory System. *By* Lloyd Hastings and True-Jenn Sun 355

Intralingual Stimulation with Sweet Proteins in Rhesus Monkey and Rat. *By* G. Hellekant, T. Roberts, C. Hård Segerstad, and H. van der Wel 356

Ultrastructure of Mouse Fungiform Taste Buds. *By* David M. Henzler and John C. Kinnamon.......................... 359

Are Aprical Membrane Ion Channels Involved in Frog Taste Transduction? Neural and Intracellular Evidence. *By* M. Scott Herness 362

Amiloride Inhibition of Neural Taste Responses to Sodium Chloride. *By* Thomas P. Hettinger and Marion E. Frank 366

Development of Amiloride Sensitivity in the Rat Peripheral Gustatory System: A Single-Fiber Analysis. *By* David L. Hill 369

Development of Olfactory and Vomeronasal Systems in the Red-Sided Garter Snake, *Thamnophis sirtalis parietalis. By* David A. Holtzman and Mimi Halpern 373

Possible Mechanisms for the Processes of Referred Taste and Retronasal Olfaction. *By* David E. Hornung and Melvin P. Enns 375

Gustatory, Thermal, and Mechanical Responses of Cells in the Nucleus Tractus Solitarius of the Frog. *By* Nobusada Ishiko, Takamitsu Hanamori, and David V. Smith 378

Steroid-Dependent Anosmia. *By* B. W. Jafek, D. T. Moran, P. M. Eller, and J. C. Rowley, III.............................. 381

Investigations of Receptive Fields of Olfactory Bulb Neurons in the Frog. *By* TAO JIANG AND ANDRÉ HOLLEY...................... 384

Tuning of Olfactory Neurons Sensitive to Hydroxy-L-Proline in the American Lobster. *By* BRUCE R. JOHNSON, CARL L. MERRILL, ROY C. OGLE, AND JELLE ATEMA.................. 388

Peripheral Coding by Graded Overlapping Reaction Spectra? Sensitivity and Selectivity of Olfactory Receptor Cells in *Antheraea polyphemus* L. *By* WOLF A. KAFKA 391

Specific L-Arginine Receptor Sites: Biochemical and Neurophysiological Studies. *By* D. LYNN KALINOSKI, BRUCE P. BRYANT, GAD SHAULSKY, AND JOSEPH G. BRAND......... 396

Cell Suspension from Porcine Olfactory Mucosa: Changes in Membrane Potential and Membrane Fluidity in Response to Various Odorants. *By* MAKOTO KASHIWAYANAGI, KIMIE SAI, AND KENZO KURIHARA.. 398

Odor Information Processing in the Olfactory Bulb: Evidence from Extracellular and Intracellular Recordings and from 2-Deoxyglucose Activity Mapping. *By* J. S. KAUER AND K. A. HAMILTON 400

Lectin-Binding Sites in Olfactory Sensilla of the Silkmoth, *Antheraea polyphemus*. *By* THOMAS A. KEIL 403

Comparisons between Human Gustatory Judgments of Square Wave Trains and Continuous Pulse Presentations of Aqueous Stimuli: Reaction Times and Magnitude Estimates. *By* STEVEN T. KELLING AND BRUCE P. HALPERN.......................... 406

Temporary Analyses of the Actions of Normal Alcohols on Taste Receptor Cell Responses to Sucrose. *By* LINDA M. KENNEDY 409

Voltage-Dependent Ionic Currents in Dissociated Mudpuppy Taste Cells. *By* SUE C. KINNAMON AND STEPHEN D. ROPER......... 413

Inhibitory Effects of Ca^{2+} on the Mg^{2+} Response of Water Fibers in the Frog Glossopharyngeal Nerve. *By* YASUYUKI KITADA... 417

Central Projections of Major Branches of the Facial Taste Nerve in the Japanese Sea Catfish. *By* S. KIYOHARA, S. YAMASHITA, T. MARUI, AND J. CAPRIO... 419

The Effects of Bilateral Sectioning of the Chorda Tympani and the Greater Superficial Petrosal Nerves and the Submaxillary and Sublingual Salivary Glands on the Daily Eating and Drinking Patterns in the Rat. *By* ROBIN F. KRIMM, MOHSSEN S. NEJAD, JAMES C. SMITH, AND LLOYD M. BEIDLER 422

Dynamic Conformational Changes of Receptor Domains in Gustatory Cell Membranes. *By* KENZO KURIHARA, MAKOTO NAKAMURA, AND YUSUKE KATOH 425

Purification and Chemical Structure of Taste Modifier: Ziziphin and Taste-Modifying Protein. *By* YOSHIE KURIHARA, KAZUYOSHI OOKUBO, AND BRUCE P. HALPERN............................. 427

Perceptual and Intraoral pH Measurements in Response to Oral Stimulation. *By* D. B. KURTZ, J. C. WALKER, J. H. REYNOLDS, D. L. ROBERTS, AND S. L. YANKELL 429

Levels of Immunoreactive β-Endorphin in Rostral and Caudal Sections of Olfactory Bulbs from Male Guinea Pigs Exposed to Odors of Conspecific Females. *By* JAY B. LABOV, YAIR KATZ, CHARLES J. WYSOCKI, GARY K. BEAUCHAMP, AND LINDA M. WYSOCKI 432

Derivation of Power Law Exponents from Olfactory Thresholds for Pure Substances. *By* PAUL LAFFORT AND FRANÇOIS PATTE.... 436

Tactile and Taste Responses in the Superior Secondary Gustatory Nucleus of the Catfish. *By* C. F. LAMB, IV AND J. CAPRIO..... 440

Histogenesis of Pontine Taste Area Neurons in the Albino Rat. *By* PHILLIP S. LASITER AND DAVID L. HILL 444

Olfaction Test in Laryngectomized Patients by the Artificial Tube Method. *By* JEUNG GWEON LEE, MASARU OHYAMA, ETSURO OBTA, KAZUYOSHI UENO, AND SHOKO KATAHIRA 447

Vomeronasal Chemoreception May Activate Reproduction in Reflex-Ovulating Prairie Voles. *By* JOHN J. LEPRI AND CHARLES J. WYSOCKI........................... 449

A Changing Density Technique to Measure Nasal Airflow Patterns. *By* DONALD A. LEOPOLD, DAVID E. HORNUNG, ROBERT L. RICHARDSON, PAUL F. KENT, MAXWELL M. MOZELL, AND STEVEN L. YOUNGENTOB 451

Electrophysiological Properties of Immature and Mature Olfactory Receptor Cells. *By* M. S. LIDOW, R. C. GESTELAND, S. J. KLEENE, AND M. T. SHIPLEY...................... 454

Decreased Sensitivity to Bitter Solutions Following Chronic Opioid Receptor Blockade. *By* WESLEY C. LYNCH, CHARLES M. PADEN, AND SUSAN KRALL...................... 456

Gustatory Activity in the NTS of Chronic Decerebrate Rats. *By* GREGORY P. MARK AND THOMAS R. SCOTT................... 458

Morphological and Behavioral Evidence for Chemoreception by Predaceous Stonefly Nymphs and Their Mayfly Prey. *By* LEE ANNE MARTÍNEZ 462

Neural Connections of the Facial and Vagal Lobes in the Japanese Sea Catfish, *Plotosus anguillaris*. *By* T. MARUI, Y. KASAHARA, J. S. KANWAL, J. CAPRIO, AND S. KIYOHARA 466

Covalent Modification of Schiff Base-Forming Proteins: *In Vitro* Evidence for Site Specificity and Behavioral Evidence for Production of Selective Hyposmia *in Vivo*. *By* J. RUSSELL MASON, LARRY CLARK, AND THOMAS HELLMAN MORTON.... 468

Skin Lipids of Garter Snakes Serve as Semiochemicals. *By* ROBERT T. MASON, JOHN W. CHINN, AND DAVID CREWS 472

Changes in Excitable Properties of Olfactory Receptor Neurons Associated with Nerve Regeneration. *By* LEONA M. MASUKAWA, BRITTA HEDLUND, AND GORDON M. SHEPHERD   475

Blocking Learned Food Aversions in Cancer Patients Receiving Chemotherapy. *By* RICHARD D. MATTES, CATHY ARNOLD, AND MARCIA BORAAS   478

Association between Anosmia and Anorexia in Cats. *By* KIMBERLY MAY   480

A Linkage between Coding of Quantity and Quality of Pheromone Gland Components by Receptor Cells of *Trichoplusia ni. By* M. S. MAYER AND R. W. MANKIN   483

Locations of Taste-Evoked Activity Are Not Coincident with Chorda Tympani Afferent Endings in the Nucleus of the Solitary Tract in the Hamster. *By* MARTHA McPHEETERS. THOMAS P. HETTINGER, MARK C. WHITEHEAD, AND MARION E. FRANK   485

Relationships between and among Selected Measures of Sweet Taste Preference and Dietary Habits. *By* D. J. MELA AND R. D. MATTES   489

A Freeze-Fracture Study on the Prenatal Development of Ciliated Surfaces in Rat Olfactory Epithelia. *By* BERT PH. M. MENCO . .   491

Mate Recognition by an Antarctic Isopod Crustacean. *By* WILLIAM C. MICHEL   494

Generalization of Conditioned Taste Aversion to Sodium Chloride in Fischer-344 and Wistar Rats. *By* ELEANOR E. MIDKIFF AND ILENE L. BERNSTEIN.   498

Human Fungiform Taste Bud Density and Distribution. *By* INGLIS J. MILLER, JR.   501

Relation of Receptive Field Size and Salt Taste Responses in Chorda Tympani Fibers during Development. *By* CHARLOTTE M. MISTRETTA, TAKATOSHI NAGAI, AND ROBERT M. BRADLEY.   504

Contribution of Cations to Generation of Salt-Induced Receptor Potential in Frog Taste Cell. *By* TAKENORI MIYAMOTO, YUKIO OKADA, AND TOSHIHIDE SATO   506

Trout Olfactory Receptors Degenerate in Response to Water-Borne Ions: A Potential Bioassay for Environmental Neurotoxicology. *By* DAVID T. MORAN, J. CARTER ROWLEY, AND GEORGE AIKEN   509

Degeneration-Regeneration of the Olfactory Neuroepithelium Following Bulbectomy: An SEM Study. *By* EDWARD E. MORRISON, PASQUALE P. C. GRAZIADEI, AND RICHARD M. COSTANZO.   512

Effects of Age and Biochemical Status on Preference for Amino Acids. *By* CLAIRE MURPHY 515

Electrophysiological and Innate Behavioral Responses of the Dog to Intravenous Application of Sweet Compounds. *By* LAWRENCE J. MYERS, RANDY BODDIE, AND KIMBERLY MAY 519

Sweet Intensity, Persistence, and Quality of Mixtures Containing Neohesperidin Dihydrochalcone and Bitter Stimuli. *By* MICHAEL NAIM, EMMANUEL DUKAN, LYAT YARON, AND URI ZEHAVI 521

Taste Responses of the Cross-Regenerated Greater Superficial Petrosal and Chorda Tympani Nerves of the Rat. *By* MOHSSEN S. NEJAD AND LLOYD M. BEIDLER 523

Genetics of the Ability to Perceive Sweetness of D-Phenylalanine in Mice. *By* YUZO NINOMIYA, TETSUICHIRO HIGASHI, TSUNEYOSHI MIZUKOSHI, AND MASAYA FUNAKOSHI 527

Axons Do Not Necessarily Compete for Targets. *By* BRUCE OAKLEY 530

Response Properties of Thalamocortical Relay Neurons Responsive to Natural Stimulation of the Oral Cavity in Rats. *By* HISASHI OGAWA AND TOMOKIYO NOMURA 532

Scanning Electron Microscopic and Histochemical Studies of Taste Buds. *By* MASARU OHYAMA, KEI OGAWA, KOZO FUKAMI, ERIKO TABUCHI, KAZUYOSHI UENO, AND KAZUYO TANAKA .. 535

Taste Concepts and Quadranomial Taste Description. *By* M. O'MAHONY, R. ISHII, AND D. SHAW 537

The Size of Mitral Cells Depends on the Age at Which Continuous Odor Exposure Commences. *By* H. PANHUBER, D. G. LAING, AND A. MACKAY-SIM 539

Brief Taste Stimuli: Cued and Uncued Magnitude Estimates and Their Reaction Times. *By* TERESA PANTZER, STEVEN T. KELLING, AND BRUCE P. HALPERN 541

Cytochrome Oxidase Staining in the Olfactory Epithelium and Bulb of Normal and Odor-Deprived Neonatal Rats. *By* P. E. PEDERSEN, G. M. SHEPHERD, AND C. A. GREER 544

Isolation and Characterization of an Odorant-Binding Protein. *By* JONATHAN PEVSNER, PAMELA B. SKLAR, AND SOLOMON H. SNYDER 547

Multiple Bitter Receptor Sites in Hamsters. *By* CARL PFAFFMANN AND M. SCOTT HERNESS 550

Zigzag Flight as a Consequence of Anemotactical Imprecision. *By* R. PREISS AND E. KRAMER 554

Putative Glutamergic and Aspartergic Cells in the Olfactory Bulb of the Rat. *By* J. L. PRICE AND T. A. FULLER 557

Benzaldehyde Binding Protein from Dog Olfactory Epithelium. *By* STEVEN PRICE AND AMY WILLEY.............................. 561

Neural Coding of Gustatory Information in the Thalamus of an Awake Primate. *By* THOMAS C. PRITCHARD, ROBERT B. HAMILTON, AND RALPH NORGREN............................ 565

Rodenticide Flavor Profiles Identified through Generalization of Conditioned Flavor Avoidance. *By* RUSSELL F. REIDINGER, J. RUSSELL MASON, AND CHARLES N. STEWARD................ 568

Specialized Receptor Villi and Basal Cells within the Taste Bud of the European Silurid Fish, *Silurus glanis* (Teleostei). *By* KLAUS REUTTER 570

Hyperinnervation Produces Inhibitory Interactions between Two Taste Nerves. *By* D. R. RIDDLE, B. OAKLEY, S. E. HUGHES, C. L. deSIBOUR, AND C. R. BELCZYNSKI....................... 574

Interactions between Taste Cells and Nerve Fibers in Murine Foliate Taste Buds. *By* SUZANNE M. ROYER AND JOHN C. KINNAMON.. 577

Postnatal Development of Cholinergic Enzymes and Acetylcholine Content in the Rat Olfactory Bulb. *By* R. SAFAEI, R. MOUSSAVI, AND E. MEISAMI.................................... 580

Bretylium Tosylate Enhances Salt Taste via Amiloride-Sensitive Pathway. *By* SUSAN S. SCHIFFMAN, SIDNEY A. SIMON, JAMES M. GILL, AND TIMOTHY G. BEEKER 584

Some Effects of Olfactory Tract Cooling on Mitral Cell Response Patterns during Repetitive Natural Stimulation. *By* D. SCHILD, T. FISCHER, AND H. P. ZIPPEL................................. 587

Contact Chemoreceptors on the Walking Legs of the Shore Crab, *Carcinus maenas. By* MANFRED SCHMIDT AND WERNER GNATZY.. 589

Further Investigations of the Receptor Potential in Primary Chemosensory Neurons by Intracellular Recording. *By* INGRID SCHMIEDEL-JAKOB, PETER A. V. ANDERSON, AND BARRY W. ACHE.. 591

Progress on the Identification of the Folate Chemoreceptor of *Paramecium. By* STEPHANIE SCHULZ, J. MICHAEL SASNER, AND JUDITH VAN HOUTEN................................... 593

Perceptual Separability and Integrality in Odor Discrimination. *By* AMY L. SCHWARTZ, MICHAEL D. RABIN, AND WILLIAM S. CAIN.. 595

Mouse Monoclonal Antibody RB-8 That Distinguishes Chemically Distinct Zones in the Primary Olfactory Projection Recognizes a 125-Kilodalton Membrane-Associated Protein. *By* JAMES E. SCHWOB AND DAVID I. GOTTLIEB 597

The Janus Head of Taste. *By* THOMAS R. SCOTT.................... 600

Chemical Analyses of Hodulcin, the Sweetness-Suppressing Principle from *Hovenia dulcis* Leaves. *By* RICHARD SEFECKA AND LINDA M. KENNEDY...................................... 602

Electron Microscopy of the Olfactory Epithelium in Zinc-Deficient Rats. *By* SHUNTARO SHIGIHARA, AKIHIRO IKUI, JUNKO SHIGIHARA, HIROSHI TOMITA, AND MASAOMI OKANO 606

Stereospecificity of the Alkyl Site for Optical Isomers of Dipeptides in the Labellar Sugar Receptor of the Fleshfly. *By* ICHIRO SHIMADA, YUJI MAKI, AND HIROSHI SUGIYAMA 610

The Role of Taste in the Feeding Mechanism of the Carp (Cyprinidae). *By* FERDINAND A. SIBBING 612

Trigeminal Chemoreceptors Cannot Discriminate between Equally Intense Odorants. *By* WAYNE L. SILVER, ADAM H. ARZT, AND J. RUSSELL MASON... 616

Differences in Epithelial Responses of Rabbit Tongue to KCl and NaCl. *By* S. A. SIMON, R. ROBB, AND J. L. GARVIN 619

The Odorant-Sensitive Adenylate Cyclase of Olfactory Receptor Cells: Differential Stimulation by Distinct Classes of Odorants. *By* PAMELA B. SKLAR, ROBERT R. H. ANHOLT, AND SOLOMON H. SNYDER... 623

Failure of Rats to Acquire a Reversal Learning Set When Trained with Taste Cues. *By* B. M. SLOTNICK, G. M. BROSVIC, AND S. R. PARKER 627

Odor Detection in Rats with Lesions of Olfactory Bulb Areas Identified Using 2-DG. *By* B. M. SLOTNICK, S. GRAHAM, D. G. LAING, AND G. A. BELL.................................... 630

Taste Bud Development in Hamster Vallate and Foliate Papillae. *By* DAVID V. SMITH AND INGLIS J. MILLER, JR. 632

Peripheral Olfactory Responses of Mature Male, Regressed, and Hypophysectomized Goldfish to a Steroidal Sex Pheromone, L-Serine, and Taurocholic Acid. *By* P. W. SORENSEN, T. J. HARA, AND N. E. STACEY 635

The Electrolyte Distribution in Insect Olfactory Sensilla as Revealed by X-ray Microanalysis. *By* R. A. STEINBRECHT AND K. ZIEROLD 638

Sequential Interactions of Oral Chemical Irritants. *By* DAVID A. STEVENS AND HARRY T. LAWLESS............................ 642

Detecting Gas Odor in Old Age. *By* JOSEPH C. STEVENS AND WILLIAM S. CAIN... 644

Voltage-Dependent Conductances in Solitary Olfactory Receptor Cells. *By* NORIYO SUZUKI 647

Multimodal Neurons in the Lamb Solitary Nucleus: Responses to Chemical, Tactile, and Thermal Stimulation of the Caudal Oral

Cavity and Epiglottis. *By* ROBERT D. SWEAZEY AND ROBERT
M. BRADLEY.. 649

Quasi-Regenerative Responses to Chemical Stimuli in *in Vivo* Taste
Cells of the Mudpuppy. *By* JOHN TEETER...................... 652

Effects of Short-Term Exposure to Lowered pH on the Behavioral
Response of Crayfishes (*Procambarus acutus* and *Orconectes
virilis*) to Chemical Stimuli. *By* ANN JANE TIERNEY AND
JELLE ATEMA.. 655

Response Characteristics of Olfactory Evoked Potentials Using
Time-Varying Filtering. *By* MITSUO TONOIKE.................. 658

Effects of Injection of Calcium, EGTA, and Cyclic Nucleotides into
the Mouse Taste Cell. *By* KEIICHI TONOSAKI AND MASAYA
FUNAKOSHI.. 662

Effect of Dietary Protein on the Taste Preference for Amino Acids
and Sodium Chloride in Rats. *By* K. TORII, K. MAWATARI,
AND Y. YUGARI.. 665

The Biochemistry of the Olfactory Purinergic System. *By* HENRY
G. TRAPIDO-ROSENTHAL, RICHARD A. GLEESON, WILLIAM
E. S. CARR, SCOTT M. LAMBERT, AND MARSHA L.
MILSTEAD.. 669

Responses of Neurons in the Nucleus of the Solitary Tract to
Lingual and Palatal Stimulation with Preferred Chemicals. *By*
SUSAN P. TRAVERS AND RALPH NORGREN 673

The Amplification Process in Olfactory Receptor Cells. *By* DIDIER
TROTIER AND PATRICK MacLEOD 677

Functional Morphology of the Olfactory Epithelium. *By*
KAZUYOSHI UENO, YUTAKA HANAMURE, JEUNG GWEON
LEE, AND MASARU OHYAMA 680

The Role of ATP and GTP in the Mediation of the Modulatory
Effects of Odorant on Olfactory Receptor Sites Functionally
Reconstituted into Planar Lipid Bilayers. *By* V. VODYANOY
AND I. VODYANOY.. 683

Receptive Fields of Second-Order Taste Neurons in Sheep:
Convergence of Afferent Input Increases During Development.
By MARK B. VOGT AND CHARLOTTE M. MISTRETTA 687

Variation in Olfactory Proteins: Evolvable Elements Encoding
Insect Behavior. *By* RICHARD G. VOGT AND GLENN D.
PRESTWICH.. 689

Signal-to-Noise Ratios and Cumulative Self-Adaptation of
Chemoreceptor Cells. *By* RAINER VOIGT AND JELLE ATEMA .. 692

Long-Lasting Effects of Context on Sweetness Evaluation. *By*
TERESA ANNE VOLLMECKE.................................... 695

Taste and Salivary Gland Dysfunction. *By* JAMES M.
WEIFFENBACH, PHILIP C. FOX, AND BRUCE J. BAUM.......... 698

Number, Size, and Density of Mitral Cells in the Olfactory Bulbs of the Northern Fulmar and Rock Dove. *By* BERNICE M. WENZEL AND ESMAIL MEISAMI 700

The Nervus Terminalis of the Shark: Influences on Ganglion Cell Activity. *By* JOEL WHITE AND MICHAEL MEREDITH 703

The Solitary Nucleus of the Hamster: Cytoarchitecture and Pontine Connections. *By* MARK C. WHITEHEAD AND LAWRENCE D. SAVOY 707

An Apparatus for the Detailed Analysis of Short-Term Taste Tests in Rats. *By* LAURA S. WILSON, JAMES C. SMITH, ROSS HENDERSON, JEFFREY SHAUGHNESSY, MOHSSEN S. NEJAD, AND LLOYD M. BEIDLER.. 710

Evidence for Participation by Calcium Ion and Cyclic AMP in Olfactory Transduction. *By* BRUCE D. WINEGAR AND ROLLIE SCHAFER.. 713

Identification of the Terminal Nerve in Two Amphibians Shown by the Localization of LHRH-like Material and AChE. *By* CELESTE R. WIRSIG AND THOMAS V. GETCHELL............... 716

Guanine Nucleotide-Binding Stimulatory Protein: A Requisite for Human Odorant Perception. *By* H. N. WRIGHT, R. S. WEINSTOCK, A. M. SPIEGEL, M. A. LEVINE, A. M. MOSES.... 719

Taste Cells in the Mudpuppy, *Necturus maculosus,* Are Electrically Coupled. *By* JIAN YANG AND STEPHEN D. ROPER.............. 723

Independence and Primacy of Umami as Compared with the Four Basic Tastes. *By* SHIZUKO YAMAGUCHI AND YASUSHI KOMATA 725

Gustatory Responses to Tetrodotoxin and Saxitoxin in Rainbow Trout (*Salmo gairdneri*) and Arctic Char (*Salvelinus alpinus*): A Possible Biological Defense Mechanism. *By* KUNIO YAMAMORI, MORITAKA NAKAMURA, AND TOSHIAKI J. HARA 727

Influence of a Single Mutation on the Incidence of Pregnancy Block in Mice. *By* KUNIO YAMAZAKI, GARY K. BEAUCHAMP, OSAMU MATSUZAKI, DONNA KUPNIEWSKI, JUDY BARD, LEWIS THOMAS, AND EDWARD A. BOYSE...................... 730

Monoclonal Antibodies Directed against Catfish Taste Receptors: Immunocytochemistry of Catfish Taste Buds. *By* JOAN YONCHEK, THOMAS E. FINGER, ROBERT H. CAGAN, AND BRUCE P. BRYANT... 732

Crustacean Chemical Perception: Tuning to Energy and Nutrient Reward? *By* RICHARD K. ZIMMER-FAUST...................... 735

Behavioral Recovery and Morphological Alterations after Lesions in the Olfactory System in the Goldfish (*Carassius auratus*). *By* H. P. ZIPPEL AND D. L. MEYER................................ 739

Index of Contributors 743

Financial assistance was received from:

- NATIONAL INSTITUTES OF HEALTH
- NATIONAL SCIENCE FOUNDATION
- BROWN & WILLIAMSON TOBACCO CORPORATION
- COCA-COLA COMPANY
- COLORADO STATE UNIVERSITY
- GENERAL FOODS CORPORATION
- GIVAUDAN CORPORATION
- PHILIP MORRIS INCORPORATED
- PROCTOR & GAMBLE COMPANY
- R. J. REYNOLDS COMPANY
- TAKASAGO CORPORATION

Preface

Every three years the International Commission on Olfaction and Taste designates a host country for ISOT, the International Symposium on Olfaction and Taste. Until 1986, the United States had not hosted this event for nearly two decades. Prior symposia had been held in Melbourne (1983), Noordwijkerhout, the Netherlands (1980), Paris (1977), Melbourne (1974), Seewiesen, West Germany (1971), New York (1968), Tokyo (1965), and Stockholm (1962). Early in 1985, the International Commission invited the Association for Chemoreception Sciences (AChemS) to host this triennial symposium in North America. Dr. Maxwell Mozell, at that time Chairman of the International Commission on Olfaction and Taste, and Dr. David Smith, then Executive Chair of AChemS, asked us whether we would be willing to organize ISOT at a convenient site in the United States. We were delighted to help out in this all-important symposium. After some investigation, we suggested that the clean, crisp air and exceptional beauty of the Colorado Rocky Mountains would be an ideal backdrop for an intensive working conference such as ISOT. An organizing committee was formed, chaired by Stephen Roper and consisting of Jelle Atema, John Caprio, Thomas Finger, Robert Gesteland, Bruce Halpern, John C. Kinnamon, Maxwell Mozell, David Smith, and Gordon Shepherd. We decided to hold the Ninth International Symposium on Olfaction and Taste at the Snowmass Village Resort, deep in the Maroon Bell Mountain Range of Colorado, July 20-24, 1986.

We felt that informal, personal interactions are the most critical components for a successful conference. We attempted to design the meeting to maximize these interactions by programming small breakfast workshops, held outdoors each day on the plaza; by including poster and slide sessions; and by leaving afternoons free for informal discussions. Even small details, such as arranging seats to be in a semicircular array in the larger conferences to facilitate discussions among the audience, spreading posters as far apart as possible to reduce interference between adjacent presentations, arranging for plentiful hot coffee and numerous comfortable chairs, were examined closely and deemed important. We were pleased to see that the 400 participants found the meeting a success and that novel research ideas and new collaborative programs have resulted directly from the interactions experienced at ISOT IX.

The organizing committee decided to divide the field of chemosensory biology into three global, but not necessarily all-encompassing questions: Where do the molecular events of chemosensory transduction take place? How is peripheral input processed in the central nervous system? Do responses to mixtures of chemical stimuli differ fundamentally from responses to pure stimuli? The symposia for the first three days focused on each of these questions in turn. The final symposium, on the fourth day, was entitled "From Reception to Perception: Summary and Synthesis."

One of the themes underlying the four-day conference was an attempt to integrate modern biophysical and molecular techniques into the study of chemosensory processes. A number of experiments utilizing patch recording techniques to study individual ion channels in receptor cell membranes were introduced at this conference. Exciting new data on molecular mechanisms of chemosensory transduction were presented. At the other extreme, novel holistic approaches to the chemical senses were presented: Two new "languages" for defining olfactory responses were described in the final symposium. Given this wide range of topics and experimental approaches, participants were exposed to an enormous menu of new ideas and new approaches to solving the questions confronting investigators. This was particularly stimulating for students and for scientists entering the field of chemical senses for the first time.

Planning an international conference involves such a vast number of helpful associates that this preface could easily become an endless litany of acknowledgments. While it is not possible to cite everyone, we would be remiss if we did not express our gratitude to Ms. Jan Chase of the University of Colorado Health Sciences Center, and Ms. Eileen Banks of the Department of Anatomy and Neurobiology, Colorado State University, for their administrative assistance in organizing the meeting. We would also like to thank Linda Mehta of the New York Academy of Sciences for her cheerful and professional assistance in seeing this book through the press.

For the organizing committee,
STEPHEN D. ROPER
JELLE ATEMA

Evolutionary Patterns in Sensory Receptors

An Exercise in Ultrastructural Paleontology[a]

DAVID T. MORAN

Rocky Mountain Taste & Smell Center
Department of Cellular and Structural Biology
University of Colorado School of Medicine
Denver, Colorado 80262

In order to survive and reproduce, all living things great and small, be they elephants or amoebae, must respond to changes in their immediate surroundings. Organisms, then, tend to be sensitive; both sensation and motility are necessary for responsiveness. Among the Protozoa, an amoeba, for example, will detect and glide away from a region of unfavorable pH. *Euglena,* armed with its photosensitive eyespot, will move toward the sunlight so essential for photosynthesis. The agile and fast-moving *Paramecium* is equipped with many motile cilia that not only serve as mechanoreceptors, but are covered by excitable membranes.[1] Among the Metazoa, an impressive array of sensory cells exists that respond to a wide range of stimuli. Investigation of the ultrastructure of metazoan sensory cells reveals a pattern of organization common to most of them: many sensory cells employ cilia, microvilli, or both at the site of stimulus reception.[2,3]

PHOTORECEPTORS

The structure of the outer segment of a "typical" vertebrate photoreceptor is drawn in FIGURE 1. Here, the stacked membranes that bear the photopigments are attached to the inner segment by a connecting cilium. Vertebrate photoreceptors, then, like many other photoreceptors, are ciliary derivatives.[4] Insect photoreceptors, on the other hand, are constructed quite differently. A cross section through a photoreceptive unit, or ommatidium, of the honeybee retina is drawn in FIGURE 2. Here, a series of retinula cells, arranged in a circle, send out microvilli—slender tubular extensions of the apical cell surface—into the center of the ommatidium.[5] Here the microvilli, which contain the photopigments, interdigitate to form the light-sensitive rhabdom. The

[a]Supported by National Institutes of Health Program Project Grant No. NS-20486 and National Science Foundation Research Grant No. BNS-821037.

1

orthogonal orientation of the microvilli is central to the bee's capacity to detect polarized light. It is interesting to note that other Metazoa, such as the annelid *Nereis,* have both microvilli and (rudimentary) cilia in their photoreceptors.[6]

MECHANORECEPTORS

A variety of ciliated and microvillar sensory cells are present in metazoan mechanoreceptors. Insect acoustic receptors, such as the ear of the noctuid moth that can detect the ultrasonic cries of bats, are innervated by chordotonal sensilla. Chordotonal sensilla, such as the one illustrated in FIGURE 3, center their function around one or two bipolar neurons, each of which sends a modified cilium to the site of stimulus reception.[7] This stands in sharp contrast to the vertebrate inner ear, in which "hair cells" are present that have an orderly array of stiff microvilli at the site of stimulus reception (FIG. 4). These microvilli, misleadingly named "stereocilia," are the cellular elements responsible for sensory transduction in inner ear hair cells.[8]

Other invertebrate mechanoreceptors, such as the insect campaniform sensillum, employ cilia that not only play a role in sensory transduction, but in development as well.[9] Each campaniform sensillum has a bipolar neuron that sends a modified cilium to a cap in the cuticle that is the site of stimulus reception (FIG. 5). Before moulting, heterometabolous insects such as cockroaches have two cuticles: an inner, "new" cuticle, which will serve as its exoskeleton after the moult, and an outer, "old" cuticle, destined to be shed at ecdysis. At this time (FIG. 6), the cilium of the campaniform sensillum undergoes tremendous elongation. It still originates at the dendrite tip of the bipolar neuron, but passes through the cap in the new cuticle, traverses the moulting space, and makes a physiologically functional connection with the cap in the old cuticle.

CHEMORECEPTORS

A variety of chemosensory cells are equipped with cilia, microvilli, or both. Insect olfactory receptors, for example, are innervated by bipolar neurons whose dendrites bear modified cilia.[10] The olfactory receptors of sharks and rays, however, employ

---

FIGURE 1. Vertebrate photoreceptor with cilium; longitudinal section through part of retinal rod.

FIGURE 2. Invertebrate photoreceptor with microvilli; cross section through ommatidium of bee eye.

FIGURE 3. Invertebrate acoustic receptor with cilium; chordotonal sensillum of insect ear.

FIGURE 4. Vertebrate acoustic receptor with microvilli (stereocilia); longitudinal section through hair cell of inner ear.

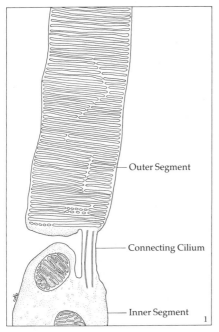

Outer Segment

Connecting Cilium

Inner Segment

1

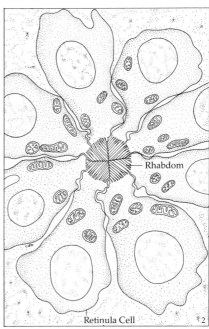

Rhabdom

Retinula Cell

2

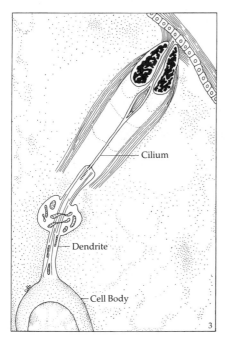

Cilium

Dendrite

Cell Body

3

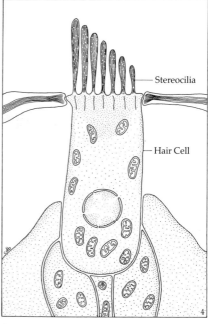

Stereocilia

Hair Cell

4

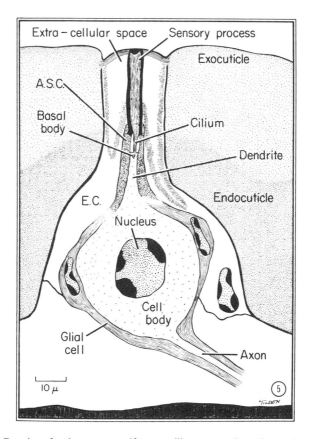

FIGURE 5. Drawing of an insect campaniform sensillum, a proprioceptive mechanoreceptor in the cuticle.

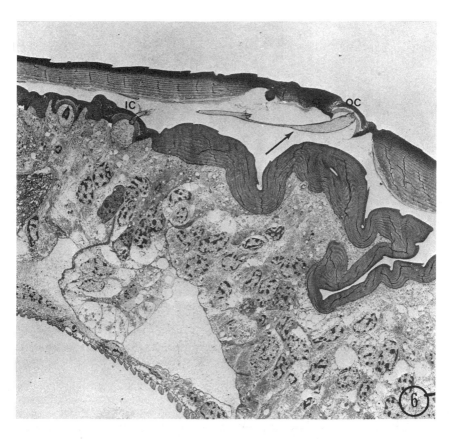

FIGURE 6. Longitudinal section through a ciliated mechanoreceptor (campaniform sensillum) of an insect before molting. Note extension of cilium (arrow) that interconnects cuticular cap in new, inner cuticle (IC) with that in old, outer cuticle (OC). Magnification × 800.

microvilli at the site of stimulus reception. Teleost fishes,[11] such as the trout, have both microvillar and ciliated olfactory receptors (FIGS. 7 and 8). Mammals, it seems, generally utilize modified cilia at the transducer sites of their olfactory receptors—and have microvillar receptor cells in the vomeronasal organs of their accessory olfactory systems.[12] Some birds, such as the duck, have olfactory receptor neurons that have both cilia and microvilli on the tip of the same dendrite.[13] Gustatory receptors tend to follow the same broad phylogenetic pattern as olfactory receptors; invertebrate contact chemoreceptors tend to be ciliary derivatives, whereas vertebrate taste cells deploy microvilli at the receptor site.

DISCUSSION

It is becoming increasingly apparent that the plasma membranes of chemosensory neurons contain specific macromolecular receptors that participate in the detection of chemical stimulants.[14-16] Cilia and microvilli, being long, slender extensions of the cell surface, provide considerable amplification of the area of the plasma membrane and glycocalyx available for stimulant-receptor interactions. Since cells, left to passively bow to minimum-energy consideration, would be spheres, it takes special intracellular structures to support axial extensions of the cell surface. Consequently, both cilia and microvilli are generated and supported by the polymerization of proteinaceous subunits into cytoskeletal structures. Cilia are supported by axonemes; axonemes are made of microtubules; and microtubules are polymers of tubulin, an ancient protein[17] as old as the mitotic spindle itself. Microvilli are supported by core filaments made of actin; actin, too, is an ancient protein, common to the vast majority of cells that can change their shape or move organelles about within their cytoplasm.

Throughout the course of evolution, sensation and motility have been closely allied to one another. The adaptive advantage of this alliance is clear; if an organism senses a change in its surroundings, its sensitivity will be selected for only if it can respond to the change by moving itself or a part of itself in response. As mentioned above, sensory cells have made widespread use of these two ancient proteins, tubulin and actin, to generate axial extensions of the cell surface that promote the molecular interactions central to sensation. Evolution tends to be conservative, and it is certainly no accident that tubulin and actin are also the basic building blocks of many major motile systems such as motile cilia, flagella, and muscle. Consequently, sensation and motility are inextricably linked together not only at the behavioral and physiological levels, but at the molecular level as well.

FIGURE 7. Scanning electron micrograph of trout olfactory epithelium showing the surfaces of ciliated (C) and microvillar (M) olfactory receptor neurons. Magnification × 11,000.

FIGURE 8. High-voltage electron micrograph of trout olfactory epithelium showing the apical poles of ciliated (C) and microvillar (M) olfactory receptor neurons. Magnification × 20,000.

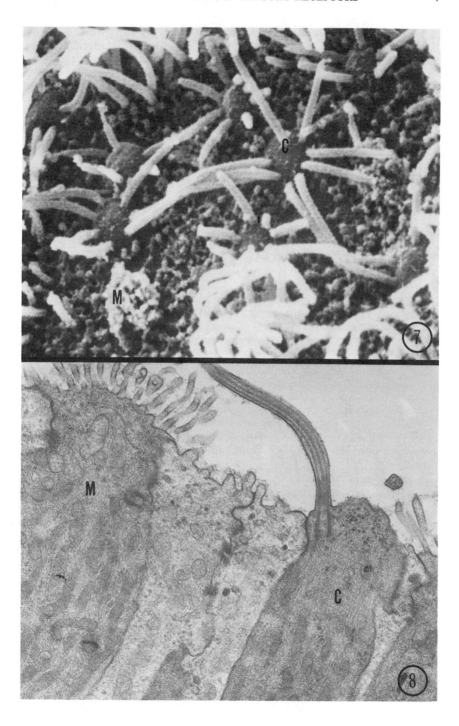

ACKNOWLEDGMENTS

The author thanks J. Carter Rowley for his collaboration during all phases of this project, Cecile Duray-Bito for making the drawings (FIGS. 1-4), Kathy Duran for assistance with photography, and George Dickel for spirited discussion.

REFERENCES

1. VAN HOUTEN, J. 1987. Eukaryotic unicells: How useful in studying chemoreception? Ann. N.Y. Acad. Sci. This volume.
2. VINNIKOV, YA. A. 1975. The evolution of olfaction and taste. In Olfaction and Taste. D. A. Denton & J. P. Coghlan, Eds. Vol. V: 175-187.
3. MORAN, D. T. & J. C. ROWLEY III. 1983. The structure and function of sensory cilia. J. Submicrosc. Cytol. 15: 157-162.
4. EAKIN, R. M. 1965. Evolution of photoreceptors. Cold Spring Harbor Symp. Quant. Biol. XXX: 363-370.
5. VARELA, F. G. & K. R. PORTER. 1969. Fine structure of the visual system of the honeybee (Apis mellifera). I. The retina. J. Ultrastruct. Res. 29: 236-259.
6. EAKIN, R. M. & J. L. BRANDENBERGER. 1985. Effects of light and dark on photoreceptors in the polychaete annelid Nereis limnicola. Cell Tissue Res. 242: 613-622.
7. MORAN, D. T., J. C. ROWLEY III & F. G. VARELA. 1975. Ultrastructure of the grasshopper proximal femoral chordotonal organ. Cell Tissue Res. 161: 445-457.
8. HUDSPETH, A. J. 1983. The hair cells of the inner ear. Sci. Am. 248: 54-64.
9. MORAN, D. T., J. C. ROWLEY III, S. N. ZILL & F. G. VARELA. 1976. The mechanism of sensory transduction in a mechanoreceptor: Functional stages in campaniform sensilla during the molting cycle. J. Cell Biol. 71: 832-847.
10. KEIL, T. A. 1986. Lectin binding sites in olfactory sensilla of the silkmoth, Antheraea polyphemus. Ann. N.Y. Acad. Sci. This volume.
11. HARA, T. J. 1975. Olfaction in fish. Prog. Neurobiol. 5: 271-335.
12. GRAZIADEI, P. P. C. 1973. The ultrastructure of vertebrates olfactory mucosa. In The Ultrastructure of Sensory Organs. I. Friedmann, Ed.: 267-305. Elsevier. New York.
13. GRAZIADEI, P. P. C. & L. H. BANNISTER. 1967. Some observations on the fine structure of the olfactory epithelium in the domestic duck. Z. Zellforsch. Mikrosk. Anat. 80: 220-232.
14. VINNIKOV, YA. A. 1986. Glycocalyx of receptor cell membranes. Chem. Senses 11: 243-258.
15. LANCET, D. 1986. Vertebrate olfactory reception. Ann. Rev. Neurosci. 9: 239-355.
16. LANCET, D. 1987. Toward a comprehensive analysis of olfactory transduction. Ann. N.Y. Acad. Sci. This volume.
17. LITTLE, M., G. KRAMMER, M. SINGHOFER-WOWRA & R. F. LUDUENA. 1986. Evolution of tubulin structure. Ann. N.Y. Acad. Sci. 406: 8-12.